BORN NEAR THE EARTH'S SURFACE
Sedimentary Rocks

Sally M. Walker

—an Earth Processes book—

ENSLOW PUBLISHERS, INC.

Bloy St. & Ramsey Ave.
Box 777
Hillside, N.J. 07205
U.S.A.

P.O. Box 38
Aldershot
Hants GU12 6BP
U.K.

Library of Congress Cataloging-In-Publication Data

Walker, Sally M.
Born Near the Earth's Surface: Sedimentary Rocks / Sally M. Walker.
p. cm. — (An Earth Processes Book)
Includes bibliographical references and index.
Summary: Describes the types of sedimentary rocks, and the processes that lead to their formation.
ISBN 0-89490-293-8
1. Rocks, Sedimentary—Juvenile literature. 2. Sedimentation and deposition—Juvenile literature. [1. Rocks, Sedimentary.]
I. Title. II. Series.
QE471.W27 1991
552'.5—dc20 90-42436
CIP
AC

Printed in the United States of America
10 9 8 7 6 5 4 3 2 1

Illustration Credits:
Courtesy Department of Geology, Northern Illinois University, pp. 16, 21, 56; J.K. Hillers, U.S. Geological Survey, p. 17; H.E. Malde, U.S. Geological Survey, p. 30; National Park Service, pp. 42, 48; S.S. Oriel, U.S. Geological Survey, p. 31; A.R. Palmer, U.S. Geological Survey, p. 23 (right); F. Peterson, U.S. Geological Survey, pp. 39, 53 (below); U.S. Department of Agriculture, p. 10; U.S. Geological Survey, p. 50; A.O. Waananen, U.S. Geological Survey, p. 32; James A. Walker, p. 55; Sally Walker, pp. 9, 12, 23 (left), 25, 35, 46, 53 (above); Stanley N. Williams, Department of Geology and Geophysics, Louisiana State University, Baton Rouge, LA , p. 6.

Cover Photograph:
James A. Walker

Contents

1

Sedimentary Rocks: Where Do They Come From?

On the morning of November 13, 1985, the town of Armero, Colombia (population 22,500) in South America, was a quiet farming community. Its residents raised crops in the rich soil of the Lagunilla River Valley. By November 14, Armero lay in ruins, and more than 22,000 of its people were dead. This disaster was the result of a lahar, or mudflow, that was triggered by the eruption of Nevado del Ruiz—a volcano about 30 miles (48 kilometers) from Armero.

The lahar that destroyed Armero began shortly before 11:00 P.M. on November 13. Nevado del Ruiz had already shot out two blasts of hot gases and ash from its snow-covered peak, and hot volcanic material began surging from the volcano. The hot material, plus heat from the ground (hot from the internal heat of the volcano), melted snow and ice on the volcano's summit. Water from the melted snow then swept up soil and loose pieces of rock and carried them downhill. This erosion—a natural process of removing rock and soil and transporting it to another area—of the flanks of Nevado del Ruiz produced at least three lahar "waves." Survivors reported that at times the mud, stone, ash, and debris-filled waves had crests of fifteen to 50 feet (5–15 meters). That's taller than most houses!

Luiz Vincente Suarez, a fruit vendor living in Armero, was lying in bed when he heard people shouting outside his house. He quickly woke his pregnant wife. Hurrying from their home, they ran toward Armero Stadium, a place they had been told would provide shelter. Outside the stadium entrance a large crowd of people scrambled and pushed, seeking safety inside. Before Mr. Suarez and his wife could enter the stadium a burning-hot lahar wave struck the crowd. The mud swept people off their feet and carried them away. Mr. Suarez grabbed for his wife, but she had already disappeared. The rapidly sliding mud carried Mr. Suarez all the way to Guayabal, a town five miles (eight kilometers) from Armero, where he finally managed to grab a tree branch. His back burned by the lahar's heat and his body coated with

Most of Armero, Colombia was destroyed by a mudflow.

thick mud, he clung to the branch and waited for help. Rescue finally arrived in the morning—five hours after he had first grabbed the branch. His wife, however, did not survive.

Rose Maria Henao and her two children were more fortunate. Mrs. Henao remembers feeling the earth tremble and the air filling with the rotten-egg smell of sulfur. She heard a horrible rumbling that seemed to come from deep inside the earth. She and her children climbed to the roof of their home located on a hill outside Armero. They saw thick slime roll into town. Houses crumbled beneath its weight. Mrs. Henao later recalled the lahar wave as moving with a moaning sound, like some sort of monster. By morning a layer of mud 7- to 15-feet (2- to 5-meters) deep had buried the central part of Armero.

Weeks later, after water had drained from the mud, a mixed layer of stones, soil, and ash remained as a 3- to 6-foot (1- to 2- meter) thick deposit. Sand, mud, and broken rock material transported and deposited in a new place is called sediment. After many years the lahar in Armero may gradually change from loose sediment to solid rock. In Utah the sediment carried by lahars that occurred thousands of years ago has changed to rock. Rock formed from sediment is called sedimentary rock.

More than half the rocks on the earth's surface are formed from sediment. Geologists—scientists who study the earth—divide sediment into two groups: clastic sediment and chemical sediment. Clastic and chemical sediments are the primary sources for the materials eventually forming sedimentary rocks. Clastic sediments are broken pieces of rocks and minerals. Beach sand and the mud on lake and river bottoms are clastic sediments. Animal remains that have been transported—sea shells and broken bones—are also considered clastic sediments. Even the dust on the furniture in your home is a clastic sediment! So, too, the deposit left by the Nevado del Ruiz lahar is a clastic sediment.

Chemical sediment forms by precipitation—the separation and settlement of particles previously dissolved in a liquid. Solids can form when chemical reactions occur in a solution, the solids precipitate and become sediment. For example, salt is a precipitate. As seawater evaporates, dissolved material in the water becomes more concentrated and precipitates as salt. Most of our table salt forms this way.

Plants and animals also produce chemical sediment. In seawater plants help decrease the acidity of the water surrounding them. The lower level of acid allows a chemical compound called calcium carbonate to form and accumulate as sediment on the seafloor. Shellfish produce calcium carbonate when they make their shells.

Weathering

There are four main steps that occur in the formation of sedimentary rock: weathering, transportation, deposition, and lithification. Weathering is the wearing down and breaking apart of existing rock. Objects exposed to water and air over long periods of time begin to decay. For example, when rain trickles around the iron body of a car it reacts chemically with the exposed metal and forms rust. Rust spots on cars are a sign of weathering. Rocks decay in similar ways.

The term *weathering* includes several different processes that act separately and together to break down rocks. These processes are divided into two main categories: mechanical weathering and chemical weathering. Mechanical weathering is the physical breaking apart of rocks. Although the rocks are broken into pieces, their chemical composition remains the same. One kind of mechanical weathering is frost wedging. In cold places water seeps into cracks in rocks and freezes. When water freezes it expands, and acting like a wedge, causes the rock to split apart. Frost wedging is responsible for the wintertime potholes that make streets bumpy.

Removal of great weight also breaks up rocks. Deeply buried rocks are under enormous pressure. When erosion removes the overlying soil and rock, the formerly buried rock is said to be unloaded. Unloaded rock tends to expand and fracture, and often pieces flake off in sheet-like layers.

Another kind of mechanical weathering is the burrowing action of insects, rodents, and earthworms, which brings small rock pieces to the surface where they are weathered. Roots can also force their way into rock crevices and cause rocks to break apart. Loose rock materials that fall on highways are usually a result of root activity and frost wedging.

Chemical weathering, the actual decomposition of rocks by dissolving the minerals that form the rock, accounts for most of the weathering that occurs. Carbon dioxide gas, found naturally in the earth's atmosphere, mixes with water and forms a mild acid called

Statue showing effects of weathering.

carbonic acid. Additional carbon dioxide from decaying plants combines with the carbonic acid and strengthens it. While not harmful to plants and animals carbonic acid (its strength is similar to soda pop left in a glass for several hours) is strong enough to slowly dissolve rocks. Calcium carbonate dissolves readily when in contact with carbonic acid. Buildings constructed with limestone, a rock made mostly of calcium carbonate, are often pitted and weathered by acids produced when industrial gases mix with rain. As the original minerals in a rock decompose, the rock falls apart.

Transportation and Deposition

During the second and third steps in sedimentary rock formation, sediment is often transported and deposited far from the area where weathering first occurs. Mudflows (like the Nevado del Ruiz lahar) and landslides move rock materials dramatically and rapidly. Fortunately sediment is usually transported more slowly by wind, water,

Air-borne particles in a dust storm during the 1930s.

and ice. The sediment carrier is referred to as the transporting agent; and after deposition, the depositing agent.

Wind carries the smallest sediment particles. During the 1930s parts of Colorado, New Mexico, Kansas, Texas, and Oklahoma became known as the Dust Bowl. Widespread farming had stripped much of the natural vegetation from the land; then several years of drought left the land very dry and dusty. Winds averaging 10 miles per hour (16 kilometers per hour) often filled the air with black dust clouds. Breathing was difficult in the sediment-filled air, and some people even died from a kind of "dust pneumonia" similar to a lung disease coal miners get from breathing rock dust.

Currents in rivers and oceans transport larger sediment grains as well as small ones. Fast-flowing streams frequently carry pebbles. As the current slows, the larger sediment settles to the bottom quickly. The tiniest particles may remain suspended for days, giving the water a cloudy appearance. In the deep ocean tiny particles may remain suspended for years. Wave action and currents along ocean shorelines transport and deposit millions of tons of sediment each year. Beaches form when sediment is washed up along a coastline by waves and shoreline currents.

Glaciers, large bodies of moving ice, are able to transport the largest sediment. They often carry boulders as big as an elephant for many miles—or kilometers. Usually sediment deposited by a glacier is a mixture of small and large rock material.

All sediment layers hold clues to the environments in which they were deposited. Ocean, swamp, river, lake, and wind deposits all have their own distinctive features.

Grain Size and Composition

Sediment particles are grouped according to individual grain size. Classifying them in this way makes it easier for geologists to understand each other when they are discussing sediments. Three sizes of particles—boulder, cobble, and pebble—are easily recognized.

Sand-size grains are also fairly easy to see and feel gritty when rubbed between your fingers. Most beaches have sand-size grains. Individual silt-size grains cannot be seen without the aid of a magnifying lens. They feel smooth between your fingers, but if you put them in your mouth they feel gritty between your teeth—like the polishing pastes used by dentists. Clay-size grains are so fine they do not feel gritty between your fingers or your teeth.

When geologists are outside examining rocks, they are referred to as being "in the field." While in the field geologists often test grain size in sediment and rocks by putting a small sample in their mouths to check for gritty texture.

In addition to classifying by size geologists also look at how well sorted—according to size—various sediment particles are. A sediment or rock with all the particles a similar size is very well sorted. If there is a mixture of large and small grains the sediment or rock is poorly sorted. In some rocks it is very easy to see if large grains are present.

This boulder, which can be rocked using a lever, was transported and deposited by ice.

A pebble-size grain or larger would be hard to miss. But in rocks made of fine-grained sediment it may be difficult to distinguish individual grains. Geologists in the field use a small magnifying lens—called a hand lens—to help them examine sediment and rock samples.

Geologists also try to identify the minerals in sediment and rock. The type of minerals present may provide information about where and how the rock or sediment formed. Three minerals (quartz, feldspar, and calcite), a group of minerals called clay minerals, and rock fragments are commonly found in sedimentary rocks.

SEDIMENT PARTICLE SIZES

Rock Name	Particle Name	Individual grain size
Conglomerate or Breccia	Boulder	Larger than 10 inches (256mm)
	Cobble	2.5-10 in. (64-256 mm)
	Pebble	0.08-2.5 in. (2-64mm)
Sandstone	Sand	0.0025-0.08 in. (0.062-2mm)
Mudrocks	Silt	.00015-0.0025 in. (0.004-0.062mm)
	Clay	smaller than 0.00015 in. (0.004mm)

Lithification

Lithification, the fourth step in the formation of sedimentary rock, is the process of consolidating loose sediment into rock. As sediment layers become deeper, the weight of the overlying sediment pushes the grains in the lowest sediment layers more tightly together. This process is known as compaction. As mineral-rich waters slowly trickle around the sediment grains, changes in temperature and pressure cause certain minerals to precipitate from the water. The precipitate cements sediment grains to one another. Thus, through compaction and cementation, loose sediment is changed to rock.

Sixty-six percent of all the surface rocks that have been mapped on the continents are sedimentary rocks. However, sedimentary rocks erode easily because they are exposed to the weathering action of water and air. Half of all sedimentary rocks are younger than 130 million years—a very short time geologically speaking.

The average thickness of sedimentary rocks on continental areas is about 5,500 feet (1,700 meters). In some areas, however, there may be little or no sedimentary rock. The Canadian Shield, a large area in Central Canada, has no sedimentary rock. In contrast, the Louisiana-Texas Gulf area has sedimentary rock layers as deep as 61,000 feet (18,500 meters).

Geologists examine many things when they study sedimentary rocks. In addition to studying grain sizes of sedimentary rock and the minerals that form them, geologists also examine sediment structures formed during deposition processes. The structures and features found inside and on the surfaces of many sedimentary rocks provide important information about the environmental conditions at the time sediments were transported and deposited.

2

Inside and On the Surface

A number of surface features help distinguish sedimentary rocks from other types of rock. One of the most common features of sedimentary rocks is the layering produced during the deposition of sediment. The layers, called beds, mark the beginning and end of a particular deposit. In cross-section, or cut in half, beds often appear as bands of color ranging from reddish to black. All beds are not the same; they vary according to the type of sediment particles they contain and the arrangement of the particles within the bed. Sedimentary rocks tend to break apart along bedding planes, the surface where one deposit ends and another begins.

Parallel Beds and Cross-Beds

Sedimentary beds can be divided into two groups: parallel beds and cross-beds. Parallel beds occur in sediment deposited on ground that is more or less level. Usually parallel beds form in watery environments where currents and wave activity are limited—the bottom of lakes and deep-sea floors, for example. Most parallel beds are fairly uniform in thickness. How thick a particular bed may be depends on the environment and the type of sediment being deposited. Sedimentary beds less than one-half inch (one centimeter) thick are called laminae. Laminated beds can be very useful to sedimentologists—scientists who study sediments—because in some cases these very thin sedimentary beds represent a single year's deposition. Laminated beds usually show a pattern of alternating light and dark layers. Lakes that receive sediment-choked water melted from glaciers frequently have this kind of lamination. The lighter, coarse-grained layers reflect the summer season when melting ice provides flowing water capable of transporting larger sediment grains. The darker layers reflect the quiet winter season when fine-grained and organic-rich material slowly settle on top of the summer layer. The decaying remains of plants and animals give the dark layers their blackish color. Each pair of light and dark layers, called a varve, equals one year's worth of deposit. By counting the varves a sedimentologist can determine the length of time it took for a particular group of sedimentary beds to be deposited.

Cross-beds, the second type of sedimentary beds, form when sediment is deposited in an inclined, or sloping, position on top of another sediment layer. Cross-bedding is the result of rapidly moving

Varves reflect seasonal cycles. Pen knife for scale.

water or air currents. This type of bedding is likely to be found in beach areas, where waves rapidly move sand particles, and in rivers where flowing currents occur regularly.

Graded Beds

By looking closely at individual parallel and cross-beds you can find clues that may provide information about the kinds of processes that occurred while the sediment was being deposited. It is important to notice the arrangement of particles within a bed if you are trying to decide how sediment might have been deposited. The particles in some sedimentary beds are very uniform, meaning their particles are all of a similar size. In other beds the sediment particles may be graded, or

Cross-bedding in sandstone.

arranged according to size, with the larger grains along the bottom of the bed and the finer grains toward the top.

To observe how graded beds form, you can do the following simple experiment. Fill an empty jar halfway with a mix of pebbles, sand, and soil. Then add water to fill the jar, put on the lid, and vigorously shake the jar. Place the jar on a table and watch what happens. The heaviest particles, the pebbles, will settle to the bottom first. Gradually, according to size, the other particles will also settle. The tiniest particles may remain suspended in the water for several hours. You will be able to see the graded layering through the sides of the jar.

For graded bedding to occur there must be a rapid and continuous loss of velocity in the way the sediment is transported. A rapid energy loss happens in several sedimentary processes. As streams lose energy, the water movement slows down. Like the heaviest sediment in the glass jar experiment, large particles quickly settle to the bottom of a stream. A sudden loss of energy in streams occurs when water recedes rapidly after a flood.

A sudden energy loss also occurs in turbid, or muddy water, usually in a lake or ocean. For example, sediment-filled water is heavier than clear water. Sometimes this layer of "heavy" water flows rapidly down a slope near a lake bottom or the seafloor below the clear water. The rapidly sliding sediment-filled water, called a turbidity current, can cause graded bedding when it reaches a level surface and slows down.

In the past, geologists who sampled sediment at the base of continental slopes (the steep underwater edges of continents) and farther along the deep-sea floor were puzzled by the graded sediment layers they found. They wondered where the larger sediment particles had come from. Strangely enough the answer—turbidity currents—has been provided by telephone and telegraph cables along continental slopes and the ocean floor.

In 1929, an earthquake occurred on the continental slope between

Nova Scotia and Newfoundland. Cables along the slope and the seafloor broke in a downhill sequence. Each time a cable snapped, the time was recorded. (Cable breaks are automatically recorded so repair ships can easily locate them.) Much later when scientists studied the records they found that the cables on the continental slope, where the earthquake had occurred, were not broken. However, cables at the base of the slope and along the seafloor had broken in sequence and at increasing depths in directions away from the earthquake area. In some places the broken cables were buried by sand and pebbles.

The movement of turbidity currents reasonably explains the breaks. Coarse sediment from the continental slope was jarred loose by the earthquake. As the sediment mixed with seawater it created a turbid layer, heavier than the clear water above. Pulled by gravity the turbulent sediment cloud flowed down the slope and across the seafloor, snapping the cables as it reached them.

By noting the times and distances apart at which the breaks occurred, scientists have calculated the turbidity current may have flowed at speeds up to 37 miles per hour (60 kilometers per hour).

Although extremely deep water makes it impossible for us to actually see the large turbidity currents, scientists have observed turbidity currents in freshwater lakes. Small turbidity currents are also often made by scientists in special laboratory tanks used to study current and particle movement.

The rapid eruption and ejection of material from a volcano can deposit sediment in a graded bed. Dust storms also deposit graded bedding. As the wind dies down, energy is lost and the particles settle to the ground according to size.

Some sedimentary beds are neither uniform nor graded. These beds are not sorted by particle size, and result from avalanches, lahars, glaciers, floating ice (icebergs), and landslides or mudflows. Like the sudden, forceful movement of the lahar at Nevado del Ruiz, these forms of transport and deposition favor a mixed jumble of particle sizes.

The amount of time it takes for a sediment bed to form varies greatly. Mudslides, floods, and landslides deposit sediment quickly—often in a matter of minutes or hours. Huge slow-moving glaciers may take hundreds of years to melt and deposit the sediment frozen inside their ice. Deep on ocean floors there are sediment beds less than one-half inch (one centimeter) thick that have taken thousands of years to accumulate.

Erosion changes the appearance of sedimentary beds. In some areas weathering processes will completely wear away sedimentary beds. In other areas the beds may be twisted and turned as the earth's crust shifts during mountain building processes.

Ripple Marks

Another common sedimentary feature is ripple marks. If you have ever looked closely at the bottom of a shallow river, lake, or bay, you may have seen ridges in the mud or sand. Perhaps you have seen them on sandy beaches or in deserts. Ripples are formed by the activity of water and wind on a sediment surface.

The circular motion of water particles and the flow of water and wind currents cause sediment particles to roll, slide, and jump along a sediment surface. This skipping kind of movement, called saltation, gradually shifts and deposits sediment into ridges. The word *saltation* comes from a Latin word meaning "to jump or leap." Small ridges, up to four inches (ten centimeters) in height, are called ripples. Larger ridges, whether deposited by water or wind, that range from 7 to 26 feet (2 to 8 meters) in height are called sand dunes. Special ocean-bottom sounding devices have detected extremely large ridges on the underwater shelves along continental edges. These giant sand waves have heights up to 49 feet (15 meters). The velocity, or speed, of the depositing agent (water or wind) and the size of the sediment particles are the most important factors that determine the size of ripples and dunes.

Ripples and dunes do not stay in the same place. Wind and water move sediment particles from one area of a ridge to another. This continual process causes ripples and dunes to migrate, or move, across a sediment surface. Dunes can move several feet (or meters) or more per year. Ripple trains, groups of parallel ripples, gradually migrate along the sediment surface in the direction of the wind or water current flow.

If you could cut a ripple in half and look at a cross-section, you would find that each ripple is made of a number of laminae. Beneath a ripple the bottom laminae are often on a level surface. The back side of a ripple is more gradually sloped than the lee, or front side. As sediment particles travel up a ripple they form a thin layer on the back of the ridge. Eventually the particles reach the crest, or top, of the ridge and tumble down the lee side. Because the lee side is a sloping surface

Lithified ripple marks.

the falling particles form cross-bed layers. The sloping cross-beds point in the direction the current is moving.

Sedimentologists are able to study ripple formation and the effects of wind and water upon them by building wind tunnels and ripple tanks in laboratories. These instruments can be set to create different environmental conditions found on the earth, and the changes caused by different conditions can be measured. By studying the present day sedimentary processes that form ripples and comparing them with lithified ripple marks, scientists can reconstruct ancient environments.

Mud Cracks

Mud cracks are a third fairly common feature of sedimentation and sedimentary rocks. Because mud is a mixture of fine-grained sediment and water (possibly as much as fifty percent), individual sediment particles are often surrounded by water and are loosely packed together. When a mud surface is exposed to air the sediment begins to dry. The sediment particles shrink together as the water evaporates from the mud surface. Cracks develop in the mud as the sediment dries and shrinks. Mud cracks are important indicators of an environment. Since mud forms in a watery place we know that the sediment was deposited in water and also that the sediment was exposed to air before turning to stone. Environments and conditions favorable to the development of mud cracks are lake bottoms as they are drying up, land near oceans where tides leave sediment exposed to air, and on muddy surfaces left behind by flooding rivers.

You may be able to find present day mud cracks in the area where you live. In hot weather, after a heavy storm, look for muddy puddles and watch what happens as the mud dries. Backyard baseball diamonds and beneath swing sets are especially good places to look.

Fossils, the hardened remains of plants and animals that lived long ago, are also a very important feature of sedimentary rocks. They occur when a plant's or an animal's remains are buried by sediment. Teeth, bones, and shells are more likely to become fossils than the soft, fleshy parts of animals because they do not decay as rapidly. After an animal or plant has been buried, the remains are gradually dissolved and replaced by minerals carried by naturally occurring chemical solutions.

Mudrocks, sandstones and conglomerates, and limestones are the most abundant types of sedimentary rock, and together account for almost ninety-five percent of all sedimentary rocks. The other five percent of sedimentary rocks includes evaporites (rocks formed by the evaporation of naturally occurring waters), cherts (hard, fine-grained rocks formed almost entirely of quartz), coal, and iron-bearing sedimentary rocks.

Sedimentary rocks can be classified by their textures, grain sizes, colors, and mineral compositions. The sedimentary processes and environments in which they form help determine what type of rock will be produced.

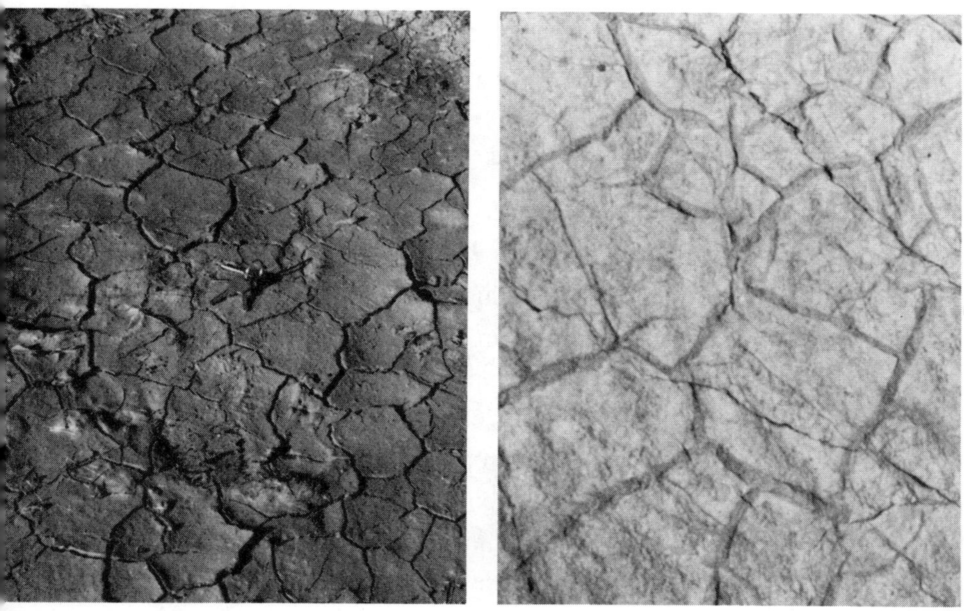

Present day mud cracks (left), and lithified mud cracks (right).

3

Mudrocks

Along the shores of the Bay of Fundy in Nova Scotia, Canada, vast mud flats are regularly covered then exposed by the tide. In some places the mud is very hard—a person can walk across its slippery surface without sinking in. In other places the clay and silt-sized sediment may be shin-deep. It isn't unusual to lose a sneaker while trying to pull a foot out of the heavy, smooth mud. In time, this muddy sediment may lithify and become a mudrock.

Characteristics and Formation of Mudrocks

Mudrocks account for about sixty-five percent of all sedimentary rocks. All mudrocks have individual grains too small to see without a magnifying lens. Clay minerals are the most abundant mineral group found in mudrocks. It is very important to understand the difference between clay-size particles and clay minerals. The term *clay-size* refers only to very tiny sediment. Clay minerals are flake-shaped minerals made of aluminum, silicon, and water. It is not uncommon to have a rock with clay-size particles that contains no clay minerals. Sedimentologists are careful to use the complete terms so no one will become confused. Quartz, feldspar (a fairly hard, glassy mineral that contains aluminum), and carbonate minerals are also commonly found in mudrocks. Most of the quartz found in mudrocks are tiny chips that have been broken off larger quartz grains during transportation. The feldspar is weathered from other rocks, particularly an igneous rock called granite. The origin of the carbonate minerals in mudrocks is uncertain. Many geologists believe the carbonate minerals are probably particles left from broken shell material. They may also have formed as a precipitate from water present during lithification.

Mudrocks also contain organic matter. One type of mudrock, called shale, contains ninety-five percent of the organic matter in sedimentary rocks. Shales are very important economically because over time the organic matter is converted into petroleum and natural gas—major sources of home heating fuel. However, the average mudrock contains a very small percentage of organic matter (less than one percent).

Mudrocks generally fall into two color groups: the gray to black group and the red-brown-green group. The colors are caused by

Tidal mud flat. Bay of Fundy, Nova Scotia.

minerals found in the sediment or by the presence of organic remains. Most mudrocks are a combination of a number of minerals and tend to be grayish or brown. If the mineral hematite, which contains iron, is present the mudrock tends to be red to brown. Clay minerals are a greenish color. The black color of some mudrocks comes from the element carbon which is present in organic remains.

Mudrocks form after fine-grained sediment has been deposited, compacted, and cemented together. Sediment deposition does not always occur at the same rate. In deep sea or very calm water sediment layers may accumulate slowly. In Alaska, sediment-filled water that melts from glaciers near the ocean deposits layers as rapidly as 30- to 60-feet (approximately 10- to 20-meters) per year! Compaction, the pressing together of sediment particles, is an extremely important part of the mudrock making process. Because mud contains so much water the flake-like clay minerals "float" about in the sediment and water mixture without any particular order. New layers of sediment are continually deposited on top of the older layers. The weight of the new layers squeezes water from the lower layers, pressing the sediment tightly together. The pressure of the overlying sediment forces the clay minerals into parallel layers. When this happens the sedimentary rock that results breaks easily along these layers. This type of mudrock is called shale.

The pressure of the overlying sediment also causes the temperature of the buried sediment to rise. After the sediment has been compacted, pressure and the higher temperature cause clay minerals to change. As they change they release the elements iron and calcium in solution. When the iron and calcium precipitate between the sediment particles, they act as the two major cements found in mudrocks.

Mudrock Environments

Environments in which mudrocks are likely to form include lake bottoms, swamps and marshy areas, flood deposits along rivers, deep-sea floors, and tidal flats—large areas of muddy sediment exposed and covered by the regular rise and fall of ocean tides. In these environments the water is generally more quiet, favoring the settling of fine-grained sediment.

Black mudrocks are good indicators of a particular environment. When a plant or animal dies its remains begin to decay. This happens because of certain chemical reactions that take place between the remains and oxygen. Normally oxygen circulates freely in water currents. However, in some watery environments there may be little or no oxygen. This is usually caused by a barrier on the bottom of a lake or sea which prevents currents from circulating. If oxygen-rich water is prevented from circulating in a body of water, the water becomes stagnant. A stagnant watery environment without oxygen is called a reducing environment. Because of the lack of oxygen, organic materials deposited in reducing environments do not decay and their remains give the surrounding sediment a black color.

Marine Mudrocks

One well-studied modern reducing environment in which fine-grained sediment is being deposited is the Black Sea in Asia. Approximately 7,000 years ago mountain building processes in the earth's crust caused a rock ridge to push up along the southwest floor of the Black Sea. This ridge, called the Bosporus Rise, reaches to within 130 feet (40 meters) of the water's surface. Oxygen-rich water from the Mediterranean Sea mixes with the surface waters of the Black Sea and circulates down to a depth of about 230 feet (70 meters). But the Bosporus is a barrier to the circulation of oxygen-rich water at deeper depths. Below 230 feet (70 meters) the oxygen content of the water decreases gradually until at 650 feet (198 meters) there is no measurable amount of oxygen in the water.

When life in the oxygen-rich water layer dies the remains settle to the oxygen-poor seafloor and do not decay. The undecayed organic matter forms black laminae on the sea bottom. These layers alternate with white laminae that contain coccoliths, the hard remains of a kind of algae.

The sediment deposited on the Black Sea floor is undergoing compaction but has not been completely lithified. The Burgess Shale in British Columbia, Canada, provides an excellent and very famous example of a reducing sedimentary environment that has been lithified. About 530 million years ago a sea existed in part of what is now British Columbia, Canada. A reef, or ridge of rock made from the secretions of marine animals, gradually built up on the sea floor. The circulation of water was limited by the gigantic reef, and like the Black Sea, the waters became stagnant. However, the sediment deposited along the edges of the stagnant basin gradually formed mudbanks. In the oxygen-rich mudbanks near the surface animal life was abundant. Occasionally, perhaps because of earthquakes or other movements in the earth's crust, the mudbanks would become unstable and portions of them would slide down the stagnant basin slopes to the deep bottom. Organisms living in the mud were swept along with the mud and buried on the seafloor—an environment where decay could not take place. The soft-bodied remains were preserved as fossils when the seafloor sediment was lithified. The Burgess Shale is an unusual and extremely rare "find." It is very seldom that scientists have the chance to see what the soft body parts of ancient life actually looked like. The rock record provided by the sedimentary process and environment of the Burgess Shale shows the widest diversity of phyla—groups of life—in all geologic history and has given geologists an extremely valuable "window" through which they may observe ancient life forms.

Mudrocks in Lake Environments

The sedimentary environments of the Black Sea and the Burgess Shale are marine, or sea, environments. Mudrocks also form in lakes.

Silt and clay-size particles are commonly found near the center of a lake's basin. In the quiet deeper water, sediments are frequently thinly laminated.

The sediment in the deep central water of Lake Constance, on the international border between Switzerland, Germany, and Austria, is laminated. The pairs of alternating light- and dark-colored layers, each only a fraction of an inch (a few millimeters) thick, are varves. The light-colored layers contain more of the mineral calcite (which is whitish) than the dark-colored layers. Scientists have calculated the rate of sediment deposit in the deep water to be about 0.2 inches (5 millimeters) per year.

When a river flows into a lake, it carries particles suspended in its current. As the stream of river water enters the lake basin, the coarse particles quickly settle to the bottom in the shallower water. Silt and clay-size particles remain suspended longer and are deposited toward the base of the slope and out into the lake.

The Rhine River is depositing sediment in Lake Constance in this way. Sediment continuously deposited by rivers flowing into lakes is chiefly responsible for the filling in of lake basins.

Playa lakes are shallow lakes that form in desert basins. They do not remain permanently filled because the hot, dry climate favors a great deal of evaporation. During dry periods it is not unusual for a playa lake to dry up almost completely. If the rainfall increases, playa lakes become larger and deeper. By studying the sediment and features found near modern playa lakes, geologists were able to "read" the record left in ancient mudrocks.

During the Eocene Period, which began about fifty-five million years ago and lasted until approximately thirty-eight million years ago, Lake Gosiute was a large lake in the southwest corner of Wyoming. It existed as a lake for approximately four million years, and its lithified sediment now forms part of the Green River Formation. A formation is a thick arrangement or sequence of rocks that are noticeably different from the rocks above and below them.

Keeping in mind the patterns observed in modern playa lake sediment deposits, geologists looked closely at the mudrocks of the Green River Formation. They noticed the mudrocks were thinly bedded, and in many places, laminated. Lamination and bedding often occur in deep water, but what helped them decide on a shallow playa environment was the large number of mud cracks also present. This meant that much of the time the mud surface must have been exposed to air drying.

The shale layers found in the Green River Formation, sometimes above and sometimes below the other mudrock layers, are called oil shales and are very rich in organic matter, another environmental clue. The organic matter in the examined shale is a kind of blue-green algae that thrives on the bottom of shallow lake environments. During dry periods the algal bed, which was a muddy ooze, dried and developed mud cracks. As the lake changed according to dry or wet climatic conditions, certain things happened to the mud cracks. Loose bits of dried algae and mud containing carbonate minerals were washed by storm waters into the cracks and also into depressions on the bottom

Playa lake basin and the surrounding mountains which are a source of sediment deposited in the basin.

of the lake basin. Over time the remains in the mud formed areas rich in hydrocarbons, more commonly called fossil fuels. The oil shale of the Green River Formation is estimated to hold the world's largest reserve of hydrocarbons.

Features found in and around the mudrock layers, plus mineral deposits present in the Green River Formation that are also characteristic of modern playa lakes, helped geologists conclude that Lake Gosiute was a playa lake.

Mudrocks in River and Wind-Deposited Environments

Mudrocks can also form from the silt and clay-size sediment deposited on a river's floodplain. A flooding river spills over its banks carrying with it an assortment of particle sizes. In fact, much of the property damage done to homes by flooding is due to the sludge carried into houses by the water. After the water recedes back inside its banks, some water may remain standing on the floodplain. Silt and clay-size particles are suspended in this water. As the water evaporates they are deposited in a sheet on the floodplain. These sheets can become very thick if the river floods on a regular basis. Before the building of the

Oil shale in Green River Formation. Shale fractures in very thin layers.

Aswan Dam in Egypt, the Nile River regularly flooded the land. For thousands of years people depended on the seasonal flooding to deposit a rich silt layer in which they would plant their crops.

Silt and clay-size sediment deposited by wind is called loess. When loess lithifies, it forms a mudrock. Loess deposits can be tens of feet thick. In the western part of China there are loess deposits more than 230 feet (70 meters) thick. The fine particles stay together very well and can form steep cliffs. Some people in the province of Gansu, China, live in homes tunnelled into loess. On December 16, 1920, an earthquake shook the area, causing loess to dislodge. The homes collapsed, and more than 100,000 people were killed.

Look around the area you live for places where fine-grained sediment is being deposited today. Examine rock outcrops—places where rocks stick out from the earth's surface. Could they be mudrocks? If they are, perhaps the area you live in was once a lake, a floodplain, or even a sea!

Sediment deposited by flooding of the Trinity River in California.

4

Sandstones, Conglomerates, and Breccias

If you were to visit Kaimu Beach on the island of Hawaii, you would most likely enjoy watching the waves, swimming, surfing, or maybe just walking along the beach. Chances are you would find this beach very different from most other beaches in one startling way: the sand on Kaimu Beach is black.

Grain Size and Composition

Most of us are used to seeing white sand beaches like those seen on the east and west coasts of the continental United States. However, all sands are not the same color. In Nova Scotia, Canada, there are beaches of dark gray sands. Bermuda, an island in the Atlantic Ocean off the coast of the southeastern United States, has beaches with pinkish sand. The color of sand depends on the type of rock materials that have weathered into the various sand grains. The black sand on Kaimu Beach, for example, was produced by lava from nearby volcanoes. When the hot lava reached the ocean the sudden change of temperature caused the lava to explode into bits. Waves continuously pounded the lava until it was reduced to sand-size grains.

Sand is sediment with diameters ranging from 0.0025 to 0.08

inches (0.062 to 2 millimeters). Sand grains are often shaped by their environment. For example beach sands are quite rounded due to the constant rolling and sorting by waves. In the desert, sand grains larger than 0.0039 inches (0.1 millimeters) in diameter are even more rounded and smooth. They often have a frosted appearance due to tiny pits on the grains' surfaces caused by other sand grains striking against them during transport by wind. River sands may be less rounded than desert and beach sands because they may not be transported as often and as rapidly.

The two most common minerals found in sandstone are quartz and feldspar. In fact, sandstones are generally classified by the amount of each mineral they contain. Quartz sandstones usually contain ninety-five percent or more quartz grains. Extremely resistant to all kinds of weathering, quartz grains last for a very long time. Many of the quartz grains in recent sandstones have been recycled over and over again from older rocks. During weathering processes the less resistant minerals in igneous, metamorphic, and older sedimentary rocks are worn away. As the quartz grains are released from their original rock they may be transported and deposited in a new place. These are commonly found in continental areas where ancient seas used to be and along large rivers such as the Mississippi and Amazon.

Although the average sandstone contains ten to fifteen percent feldspar grains, some sandstones contain up to fifty percent feldspar. These sandstones, called arkoses, are less durable than quartz sandstones because feldspar breaks and dissolves more easily during the weathering process. For this reason feldspar grains are not recycled as many times as quartz grains. Most likely the sediment in arkoses was transported for a relatively short distance and then buried fairly rapidly before weathering processes had a chance to wear away the feldspar grains. Mountainous areas—for example the east side of the Rockies—and river environments are places where feldspar is likely to be buried rapidly. The sandy sediment on beaches and in deserts is pounded and tossed by waves and wind too frequently for feldspars to remain as sand-size particles.

Another type of sandstone is formed when a number of small rock fragments are cemented together. Sandstones of this type, called lithic sandstones, sometimes have rock fragments that make it possible to determine exactly what type of rock the sediment came from. For example, a small piece of basalt, which is a type of lava, positively points to a volcanic source. The kinds of rock fragments found in lithic sandstones will depend on how well the fragments are able to withstand weathering and transport. Fragments of older sedimentary rocks are not often found in lithic sandstones because the cements holding the grains together easily disintegrate during weathering and transport.

Conglomerates and Breccias

Sediment with mixed grain sizes, most of which are larger than

Unconsolidated gravel.

sand-size, is called gravel. When gravel lithifies and the large pebbles are rounded, the rock is called a conglomerate. The pebbles are rounded before lithification by the transporting agent. For example, the rolling action of water often rounds pebbles. Rocks transported by glaciers may also be rounded if they are ground against other rocks during transport. If the large pieces are angular the rock is called a breccia. Breccias form at the base of a slope where loose rock has tumbled to the ground, in glaciers when the fragments have been frozen in ice and not scraped and rounded by other rocks, and in unstable areas where mountain building processes and earthquakes regularly dislodge rock materials. Their rapid deposition and burial prevents rounding.

Because conglomerates and breccias contain large fragments of other rocks, their colors also vary widely. Often the pebbles in a conglomerate or a breccia are a different color and type of rock from the matrix, the smaller sediment particles that surround the larger pieces and often hold them together.

Lithification

Sands and gravels must be compacted and cemented to become sandstones, conglomerates, and breccias.

Sand does not contain as much water between individual grains as the clay-sized particles in mud. Because of their weight and shape sand grains are more tightly packed together. If you've walked in wet sand you may have noticed that it feels much harder under foot than mud. The sand grains actually scrunch against one another as you step on them.

Some petrologists, geologists who study rocks, wondered whether sandstones, conglomerates, or breccias were likely to form if gravels were compacted without cementation. To answer their question they attempted to make a sandstone.

They began by filling a metal cylinder with a mixture of 100 percent loose quartz grains and water. One end of the cylinder was covered by a screen with very tiny holes to allow water from the sand to drain out. Next they used a piston, a short cylinder that fit tightly inside the other cylinder, to press the quartz grains together. The piston was pushed by a machine capable of applying tremendous pressure. Using pressure equal to that found in nature, they attempted to get the quartz grains to lithify. After a certain length of time they opened the cylinder. The quartz grains simply poured out.

They tried the experiment again but changed the sediment to a mixture of eighty percent quartz grains and twenty percent mud. After pressure was applied and removed, some of the grains had begun to lump together, but the result was not a complete rock. When the lumps were examined under a microscope it was found that some of the quartz grains had become embedded in the flake-shaped mud particles.

The experiment was tried a third time using 100 percent mud fragments, schist (another rock containing many flake-shaped particles) fragments, and water. This time the sediment completely lithified into a cylinder-shaped mudrock. The petrologists concluded sands of 100 percent quartz grains were unlikely to lithify solely by compaction.

For lithification of sands and gravels to occur a mineral cement is usually necessary. The "cements" form when mineral solutions carried in groundwater seep into the tiny spaces, called pores, and precipitate between sediment particles. The pressure of sediment layers as they become deeper causes the temperature around the buried sediment to rise. Many mineral solutions change as their temperature increases: some minerals will precipitate while others can exchange atoms with the sediment particles. The individual sand grains are slowly cemented together. When the pores in sediment are completely filled by a cement a very hard rock is formed.

Sandstones with moderately well-cemented spaces are also hard rocks, but because all the pores are not filled, water and other solutions are able to seep through the rock. The rock is said to be porous. Wells are drilled into porous sandstones because vast quantities of slowly flowing water are contained inside the rock. Petroleum and natural gas are also found in porous sandstone. Sandstones with poorly cemented grains are easy to break apart. Weakly cemented sandstones can be broken by pushing hard on them with a finger.

Quartz and calcium carbonate solutions are common cements. Quartz is a very durable cement and resists weathering, while calcium carbonate is less resistant to weathering—particularly when acids are present. To examine a sandstone cemented with calcium carbonate petrologists often dissolve the cement with hydrochloric acid. This procedure frees the sediment grains for individual study under a microscope.

Hematite, an iron-bearing mineral, is another common cement found in sandstones and conglomerates. When hematite cement is present the resulting rock is often orange to red. Along the eastern coast of the United States, particularly in North Carolina and southern New Jersey, the presence of hematite has colored many inland sands bright orange. Many of the sandstones in the western United States are colored reddish by hematite cement.

Sandstone, Conglomerate, and Breccia Environments

Sands and gravels are being deposited in many different environments today. The most obvious places are lake and ocean coastlines, where they form beaches. However, not all beaches are sandy. Some beaches are a gravelly mixture of sand and pebbles. Along the shoreline of Lake Superior there are beaches totally made of well-rounded cobble-sized rocks. Sand is also deposited in deltas at the mouths of large rivers. Other places are in the world's deserts, where sand is constantly being shifted by the wind. Glaciers also constantly move and deposit sediment, which in time, may become conglomerates. Once sand and gravels are deposited, they may begin the compacting and cementing processes that form sandstone, conglomerate, and breccia.

The Navajo Sandstone, a quartz sandstone named after the Navajo Indian Reservation in Arizona and Utah, is found in many outcrops throughout south and southeastern Utah and northeastern Arizona. It also extends into northern Utah and Wyoming as well as southern Nevada—where it is called by other names. Study of this widespread ancient sandstone has lead to several conclusions about the sedimentary processes under which it formed.

The Navajo Sandstone is characterized by layer upon layer of cross-bedded deposits that vary in thickness from a few inches to many tens of feet. There are occasional horizontal beds in between the cross beds. In Zion National Park, Utah, the Navajo Sandstone at its thickest measures 2,300 feet (701 meters). The average thickness in other places ranges from 300 to 800 feet (91 to 244 meters). It is generally accepted that the sediment forming the Navajo was deposited during the Jurassic Period, a geologic time period spanning from 138 to 205 million years ago.

In the field geologists used hand lenses to examine the Navajo Sandstone. They noticed the sand grains were fine- to medium-size

Navajo Sandstone.

and were very well sorted. They examined and measured the cross-beds and also found ripple marks on the back side of the bedding similar to the ripple marks found on the backs of modern sand dunes. Ventifacts, rocks that have been shaped and polished by wind, were also present. It was concluded that the ancient environment where these large cross-bedded, wind-worked sands existed was an extensive arid dune field. One of the reasons they believe the climate was mostly arid is because modern large-scale dune fields are, for the most part, barren of vegetation. When vegetation is present large dunes do not shift as readily since plant roots tend to hold the sand grains in position.

When the horizontal layers were examined, geologists found two kinds of ripple marks. The uneven ridges of one kind of ripple mark indicated flowing water that moved in a westward direction, while the other kind was very even and regular like those found today in shallow standing water. They also found burrows of small creatures that lived in pond-like environments and the tracks of three-toed carnivorous dinosaurs which were land-dwelling creatures. There was also evidence of former mud cracks, leading the geologists to conclude that the water periodically dried up, leaving only small ponds of water scattered throughout the basin.

Sand is used in the manufacturing of bricks, concrete, mortar, plaster, and of course, sandpaper. The glass industry also uses sand. Sandstone has often been used as a building stone. The famous brownstone houses in New York City are built of sandstone.

The next time you go to a city take a close look at the buildings around you. See if you can find any constructed of sandstone. As you walk, remember that the sidewalks were once loose sand produced by weathering and later transported and deposited by sedimentary processes.

5

Limestones

Every evening during the summertime a black cloud billows above the ground in an area in southeastern New Mexico. It isn't a rain cloud, nor is it smog. The cloud is actually thousands of Mexican free-tailed bats flying out of the entrance of Carlsbad Caverns. Carlsbad Caverns National Park, a series of underground chambers, contains one of the world's largest caves, covering more than fourteen acres. Caves are just one of the many interesting features found in another kind of sedimentary rock called limestone.

Grain Size and Composition

Limestones are different from mudrocks, sandstones, and conglomerates in several ways. Unlike the other rocks, the sediment and mud that eventually become limestone are formed primarily in the basin where they are found. If the sediment has been transported it is usually only a short distance from the source. Limestones are formed predominantly by the activity of aquatic creatures such as algae, shellfish, and coral, and by chemical reactions that occur between organisms and water. Most limestones are marine in origin. Limestones also differ in their mineral composition, containing mainly carbonate minerals—most frequently calcite, dolomite, and aragonite.

The remainder is usually quartz and/or clay minerals that have been transported into the sedimentary basin where limestones are forming or chert, a form of quartz. Limestones can be readily identified in the field by a simple test. Limestone fizzes when a drop of cold, dilute hydrochloric or sulfuric acid is placed on it. The fizzing is caused by the reaction of the acid with the mineral calcite.

The variably-sized sediment particles that form limestones are usually divided into four groups: ooliths, pellets, fossils and skeletal remains, and clay-sized carbonate sediment called micrite.

Ooliths are ball-shaped grains less than 0.08 inches (2 millimeters) in diameter—most are 0.008 to 0.002 inches (0.2 to 0.5 millimeters). Geologists believe ooliths form in two ways. In warm waters supersaturated with calcium carbonate, layers of calcium carbonate precipitate around a shell fragment or quartz grain as it is rolled by waves and water currents. A second way ooliths form is if a sticky

Carlsbad Caverns.

gelatin-like substance produced by seaweed coats a shell fragment or quartz grain. Bacteria living in the coating may cause chemical reactions around the fragment, which could lead to the precipitation and formation of the calcium carbonate layering. Some ooliths have only one calcium carbonate layer while others have several. Cut in half and viewed under a microscope ooliths appear as concentric circles around an unlayered nucleus.

Environments in which ooliths are likely to develop include oceans, lagoons, lakes, and tidal flats where waves and water currents keep the water agitated. Since ooliths generally form in moving water environments rather than still ones, oolith deposits are often cross-bedded.

Pellets, a second type of sediment particles found in limestone, are the round- to oval-shaped feces (or solid material waste) of small marine creatures. Marine worms and arthropods—crabs and lobsters, for example—produce pellets as they eat their way along the bottom. Unlike ooliths, pellets are not layered around a shell fragment or quartz grain. Pellets range from coarse silt-size to fine sand-size.

Fossils and skeletal remains, the third type of sediment particles found in limestones, are widely varied in size and appearance, according to the sort of creature. The complete skeleton, or more frequently, the broken pieces of a skeleton, are considered sediment particles. Skeletons of microscopic creatures become a fine-grained limestone. Larger marine organisms such as clams, oysters, and snails make coarse-grained limestone.

The fourth kind of sediment particles found in limestones are called micrite. These extremely small crystals of calcium carbonate can only be seen with very powerful microscopes. In fact, the name micrite is shortened from "microcrystalline calcite." Micrite grains are needle-shaped crystals and range in color from white to black, depending on what other materials are mixed in with them. Sediment made almost entirely of micrite is called carbonate mud. Most modern carbonate mud contains crystals of aragonite, and most of these crystals are produced by the disintegration of certain kinds of green algae.

Algae can be roughly divided into two groups: those containing no hard parts—blue-green algae, for example—and algae containing an internal structure made of aragonite needles. When algae with hard parts decay, the tissue surrounding the hard part dissolves and the aragonite needles settle to the bottom. The micrite in ancient limestones are made of calcite and their shape has usually been altered from needle-like to rice grain-like. Geologists believe that the needles were actually aragonite when they first became sediment, but because of the unstable nature of aragonite during the rock-making process, they converted to calcite—a much more stable mineral.

Shell fragments that have been ground very fine by weathering and transport are another source of micrite in carbonate mud. It is also possible that some micrite are formed by the precipitation of aragonite directly from aragonite-saturated seawater. Carbonate mud is being deposited now in many environments such as lagoons, tidal flats, shallow- and deep-sea floors, and lakes.

Algal Mats

Aquatic plants also contribute to the formation of limestone. By observing sediment patterns deposited by modern algae, geologists have been able to explain how certain kinds of ancient carbonate structures formed. On shallow, warm tidal flats and in warm fresh- and salt-water lakes and marshes, blue-green algae clump together and form large algal mats. These mats hover along the sediment surface and actually penetrate the top few millimeters of sediment. But the tissue of the algae is so fine the sediment surface is not colored green. The Bahamas, Florida, the Arabian Gulf, and Shark Bay in Western Australia are several areas where this type of algal mat is found today. Algal mats, although not carbonate material themselves, lead to the development of limestone in an interesting way. Blue-green algae have a twisted, thread-like structure somewhat similar to a basket weave. Carbonate sediment particles get trapped in the spaces between the blue-green algae's organic threads. The algae grow up and around the sediment. The new growth catches more carbonate sediment and the algal mat becomes thicker. When the algae die the remains decay and the trapped carbonate particles settle to the bottom forming laminated layers. Over time the layers lithify. Masses formed of such layers, called stromatolites, are present throughout the geologic record and in some areas form limestone deposits many feet thick. Stromatolites have been found in the Green River Formation and were one of the criteria used in formulating the Lake Gosiute reconstruction.

Lithification and Varieties of Limestone

Two main processes are responsible for the lithification of carbonate sediments: cementation and neomorphism. Aragonite and calcite are common cements. Like the cements in mudrocks and sandstones, the cements in limestones are carried in solution and precipitate in pore spaces between sediment particles.

Neomorphism is another process during which carbonate sediment is lithified. The word *neomorphism* comes from two Greek words: *neos* and *morph*, which mean "new" and "form" respectively. Since limestones form in wet environments the changes that take place are a result of a mineral dissolving into solution and then precipitating again as another carbonate mineral. Carbonate sediment undergoing neomorphism becomes a different carbonate mineral when the arrangement of atoms within the sediment particles is changed.

During lithification aragonite is often replaced by calcite. When calcite dissolves it recrystallizes as calcite again. The carbonate minerals that form limestones are soluble at lower temperatures and pressures than the minerals commonly forming mudrocks and sandstones. For this reason carbonate sediment does not have to be buried as deeply or as long before it begins to lithify. Lithification can occur in the weathering zone close to the earth's surface. Partially lithified limestones have been found with beer cans incorporated into the sediment—definitely a recent development.

There are several different varieties of limestone. Coquina is a limestone formed when shells are cemented together. The oldest continuously settled city in the United States is St. Augustine, Florida, founded by the Spanish as a military outpost in 1565. The early settlers built their fort and many of their homes with carved blocks of coquina.

Chalk is a variety of limestone composed of coccoliths and foraminifera (microscopic fossils). Chalk is soft enough that you can scrape it apart with your fingernail. The largest deposits of chalk formed on sea bottoms during the Cretaceous Period, a geologic time period lasting from approximately 135 to 70 million years ago. In fact, the Cretaceous Period got its name from the Latin word *creta*, meaning "chalk." The White Cliffs of Dover in Great Britain are chalk beds deposited during the Cretaceous Period and then uplifted years later during mountain building. Across the English Channel corresponding

Coquina.

chalk beds are found along the coast of France. Chalk deposits in the United States are found in Arkansas, Louisiana, Texas, and Wyoming.

Like mudrocks and sandstones, limestones can also have parallel bedding, cross-beds, graded beds, ripple marks, and mud cracks. But because carbonate sediment dissolves and precipitates during the rock-making process these structures and features are often erased.

Reefs

Mudrocks and sandstone form in low-lying areas when sediment particles are deposited. Limestones do not necessarily have to form in low areas. Reefs are built up from the basin floor by marine organisms and may be many tens of feet thick. Because reefs are constructed by living organisms they have sometimes been referred to as "living rock," but only the surface of the deposit is "alive." In modern seas, reefs form in warm tropical waters and in warm temperate climates where great swings to low temperatures don't occur. Reefs are found in such marine waters around the world. The Great Barrier Reef, along the east coast of Australia, is a famous modern reef. It is almost 1,000 miles (1,600 kilometers) long.

Most likely ancient reefs formed in environments similar to those of modern reefs. In some areas ancient reefs indicate that the climate and environment were vastly different from the present. The land in southeastern New Mexico, now a semi-arid plateau, was once a sea about the size of the present day Black Sea. During the Permian Period, a geologic time period about 250 million years ago, a large reef built by sponge, bryozoan (tiny invertebrate animals that live in colonies), and algal remains formed along the perimeter of the sea. It lithified into a fine-grained, unbedded limestone formation. At the end of the Permian Period the sea level was lowered during a time of worldwide glaciation.

The reef, now called the Capitan Limestone, was exposed to the weathering processes of surface water solutions and groundwater. In later times the sea underwent long periods of evaporation, and a mineral called gypsum, which forms when evaporation occurs, was deposited alongside and around the reef. In many places the gypsum deposits are as high as the reef. Several million years later the whole region was uplifted during a time of mountain building.

Sulfuric acid, produced by chemical reactions that occurred during the mountain-building process, and carbonic acid, found in the soil and in rain, gradually mixed with water inside the ground. Pulled

El Capitan limestone reef just above the surrounding land.

downward by gravity, the acidic water seeped through cracks and pore spaces in the rock, slowly dissolving the limestone. For several more million years, acidic water continued to form large underground cavities filled with water. Sometime within the last two million years the land was uplifted again and the water drained from the rock. At least 20 miles (32 kilometers) of underground rooms and passageways lay hidden beneath the earth's surface. Later, smaller amounts of mineral-rich water seeped through the rock and precipitated inside the cavern, creating the dramatic rock formation we now call the Carlsbad Caverns.

Sinkholes and Karst

Another interesting land feature created in areas underlain by limestone formations is the sinkhole. A sinkhole is a large pit. It forms in much the same way as a cave. In fact, some sinkholes were caves at one time. Their roofs collapsed, exposing the caves to the open. Many sinkholes drain into caves. Regions where the texture of the land is produced by solution are called karst. Karst topography is named after a plateau in Yugoslavia where this particular landscape is well developed. Parts of Florida, Kentucky, Tennessee, Pennsylvania, and Indiana are some of the places where sinkholes and karst topography are found in the United States.

Sinkholes can be dry or wet. In Mexico, in the Yucatan Peninsula, there are many sinkholes partially filled with water. The ancient Maya used them as a source of water. The bottoms of these sinkholes always hold some water. In Chichen Itza, a large Mayan city, one sinkhole was believed to be a sacred place. The Maya dedicated it to their rain gods. Many pieces of copper, jade, and gold have been found in the bottom of the sinkhole, along with the bones of human sacrifices, thrown in as offerings to the gods.

In regions where carbonate rocks are widespread and thick many sinkholes and caverns may develop. Extensive dissolution of flat-lying carbonate bedrock causes steep-sided hills to develop. Continued dissolution leaves the hills as towers of rock. In southern China along the banks of the Li River these towers have almost the appearance of gigantic statues huddled on the riverbanks. This kind of topography is known as tower karst.

Limestones are an important type of rock economically. They are the chief source of lime, a substance that forms when limestone is burned. Lime is used in making cement and mortar. Farmers neutralize acidic soil by adding ground limestone to the soil. Crushed limestone is often used as a base gravel for macadam roads. Limestone is also widely used as a building stone. Limestones from Indiana composed of ooliths are an especially fine building stone. Chalk is also widely used in industry. In the past, most chalk used for writing on blackboards was natural chalk shaped into stick form. Today, blackboard chalk is made from gypsum. However, the colored pastels used by artists are still often made with natural chalk. Ground-up chalk is used as a soft polishing powder and is also used in tooth powders.

This sinkhole in Bartow, Florida, was 520 feet long, 125 feet wide, and 60 feet deep.

6

Other Important Sedimentary Rocks

Mudrocks, sandstones, and limestones account for about ninety-five percent of all sedimentary rocks. However, the remaining five percent includes some of the sedimentary rocks most economically important to people. Sedimentary processes are crucial to the formation of chert, coal, gypsum, and large iron-bearing rock deposits. Today, as well as in the past, these rocks have affected human lifestyles in many ways.

Thousands of years ago many people depended on well-crafted, sharp spearpoints and arrowheads for their survival. Chert was one rock they used to fashion the points.

Chert

Chert is primarily composed of a mineral called chalcedony, a translucent structureless form of quartz. Usually white to light gray in color, the freshly broken surface of chert has a waxy shine. Chert is generally formed by the skeleton-producing activity of certain marine and fresh-water microorganisms. These particular animals and algae remove dissolved silica from water. The silica precipitates inside the organism forming a solid skeleton. After the animal or plant dies, its remains sink to the ocean floor becoming a loose microscopic sediment called an ooze. When a sufficient quantity of skeletons accumulate, changes in temperature and pressure within the ooze cause the silica to recrystallize as chert.

In non-marine water, a lake for example, chert may form if the pH, or acid level, of the water is high. Highly acidic water dissolves more silica into solution. At a later time the pH of the water may be lowered, perhaps during seasons of heavy rain. The extra silica precipitates as a gel along the lake bottom and may crystallize as chert.

Chert is found in two forms: bedded chert and nodules. Bedded cherts vary in thickness with individual beds ranging from an inch to tens of feet. Ancient bedded cherts are found all over the world. Those formed within the last 570 million years are mostly marine in origin and are the result of the organic activity of the microorganisms discussed earlier. In modern times oozes that may eventually form bedded cherts are primarily found on ocean floors and often cover extensive areas. Occasionally chert beds show cross-bedding or ripple marks, but most frequently bedding is the only sedimentary structure.

Nodules, ball-shaped pieces of chert, are commonly found embedded in limestone. This happens when the silica skeletons or silica gel are buried by carbonate sediment. The silica recrystallizes, becoming chert, and the surrounding carbonate sediment lithifies into limestone.

Gypsum and the Formation of Evaporites

Gypsum is the most commonly found mineral in sedimentary rocks called evaporites. Evaporites precipitate when marine- and freshwaters evaporate. There are more than 70 different evaporite minerals. Evaporites are very soluble so they aren't found in extensive outcrops: rain and water flowing within rocks beneath the earth's surface dissolve them. Gypsum, the least soluble of the evaporites, does occur in outcrops, but it is still a soft mineral and can easily be scratched with a fingernail.

Although evaporites don't cover large areas of the earth's surface,

Spear points and arrowheads made of chert (above) and chert nodules in limestone (below).

they are present in very large quantities just below the surface. In western Texas and southeastern New Mexico, the Castile Formation, composed primarily of the evaporite minerals gypsum, anhydrite, and halite (rock salt), covers more than 40,000 square miles (64,000 square kilometers) and is 3,960 feet thick (1,200 meters) in some places. Although there are some gypsum outcrops, most of the Castile Formation is beneath the land surface. This huge formation formed in the same sea basin as the Capitan Limestone.

The formation of evaporites depends on a dry climate where evaporation exceeds precipitation. Therefore, certain areas of the earth are more favorable to evaporite formation than others. Air circulation patterns in the earth's atmosphere play an important role in determining where dry climates will form.

The earth's surface is closest to the sun along the equator. Warm air, heated by the sun, rises into the atmosphere, carrying moisture along with it. The moisture filled air circulates north and south, away from the equator. The air cools as it rises and water droplets condense. When the water droplets become heavy enough they fall to the earth as rain. The large amount of rainfall plus the warm air temperature create a suitable climate for the growth of tropical rain forests in equatorial regions. After the air loses its moisture it continues to spread north and south. By the time it reaches about 30 degrees latitude, there is very little or no moisture left that can precipitate as rain. The dry air absorbs moisture from the land surface. As a result, the land becomes arid and a desert develops. Most modern deserts (the Sahara Desert in Africa, for example) are located in regions about 30 degrees latitude north or south of the equator. The North and South Poles are also arid because cold air does not carry much water. In Antarctica there are many cold evaporite lakes.

Evaporites form in two ways. If a sea or lake no longer receives water and evaporation occurs, evaporite minerals may precipitate. They also precipitate in the pores inside some rocks and between the sediment particles in soil layers on or near the earth's surface.

Evaporites that develop inside soil layers form a crusty cover in a desert area.

The Dead Sea, a large lake between Jordan and Israel, is one area where evaporites are forming at present. This lake, more than six times saltier than the ocean, has no fish or any other animals living in its waters. The area surrounding the lake has scarcely any vegetation. The Great Salt Lake in Utah is another basin in which evaporites are presently forming.

Evaporite beds are important economically. Rock salt, used to de-ice roads in winter, and table salt are excavated from mines dug into evaporite beds. Salt domes—dome-shaped sedimentary rock formations with rock salt at their core—are used for storage of harmful chemicals and may be sites for radioactive waste disposal.

Powdered gypsum, the main ingredient used to make plaster and

Large plain encrusted with gypsum.

wallboard for home and office construction, is also excavated from evaporite beds.

Coal

Deep inside winding tunnels people dig another type of rock from the walls around them. They are mining coal. Coal, used to heat many homes, is a unique variety of sedimentary rock. Unlike other sedimentary rocks, it is formed from deposits of land plants.

One of the most important conditions necessary for the deposits to develop into coal is a reducing environment. In most soils there is sufficient oxygen present for plant decay to take place. But in swampy areas—bogs and moors for example—the plant remains are often deposited in shallow, stagnant waters.

When a flower is pressed between the pages of a book it is quite flat. Organic remains are even thinner, so it takes many years of very lush growth to produce thick layers of pressed plant remains. The plant layers accumulate and are eventually buried by more sediment, usually sand or mud. Hydrogen, oxygen, and carbon are the three major

Seam of black coal.

elements that make up the layered plant remains. Layers of sand and mud apply pressure to the plant remains and the temperature rises. The increased pressure and temperature cause chemical reactions to occur which gradually drive off the oxygen and hydrogen, leaving mostly carbon behind.

After burial the woody plant tissues begin to clump together and form a soft, brown material called peat. Peat, which is about sixty percent carbon and thirty percent oxygen, can be crumbled with your fingers. Peat is easily lighted and burns with a smoky flame. As the process of converting peat into coal continues, it is not unusual for methane gas to be given off. When methane gas, also called swamp gas, ignites it produces an eerie glow in marshy places that has often been reported as UFO sightings or ghosts.

If peat remains buried for a longer period of time, more oxygen and hydrogen are driven off and a low-grade coal, called lignite, forms. The carbon content of lignite is about seventy percent. Lignite can still be broken apart with your fingers.

Continued burial finally yields a hard coal called anthracite, which is ninety to ninety-five percent carbon. This shiny, black coal is harder to ignite than peat or lignite, but once ignited it burns with a smokeless flame. Generally speaking the longer the plant remains are buried, the harder and better the grade of coal produced. Because it has been changed so much by temperature and pressure, anthracite is considered a metamorphic, rather than a sedimentary, rock. If anthracite remains buried long enough, the high temperatures eventually transform it into graphite, the mineral used to make pencil "leads."

During its formation coal gives off various gases, some of which are highly combustible. Sometimes fires begin in abandoned coal mines. Once a coal mine fire begins it is very difficult to get under control. Some fires will burn in mine tunnels for many years, until they burn themselves out.

Ancient deposits of coal are found in many places all over the world, including the Appalachian Mountains in the eastern United

States, a number of places in the western and midwestern United States, England, Scotland, the Soviet Union, and China.

One of the many present environments where peat is being produced is the Great Dismal Swamp, in North Carolina and Virginia.

Ironstones and Iron Formations

Iron-bearing sedimentary rock is particularly important economically because of the iron ore removed from the rock. Without iron ore our lives would be vastly different—no cars, planes, skyscrapers, or trains.

The earth's major iron deposits are found in ironstones and iron formations. Although almost all sedimentary rocks contain some iron, the iron-rich ironstones and iron formations contain more than fifteen percent iron. The iron in some ores may be as high as thirty-five to forty percent. The oldest iron-bearing sedimentary rocks—the iron formations—formed earlier than 570 million years ago, while ironstones have formed since then. Geologists divide the iron-rich sedimentary rocks in this way because the older rocks differ from the younger rocks in several ways. Iron formations are very extensive in area, some deposits can be traced over hundreds of miles. They are thinly bedded and often are layered with chert. The thickness of the entire deposit can range from 165 to 2,000 feet (50 to 600 meters). Calcite is seldom found in iron formations. The ironstones, on the other hand, seem to have formed in basins that didn't cover as much area. Ironstone deposits are usually only a few feet to a few tens of feet thick. Fossils are frequently found in ironstones, but not at all in iron formations. Ooliths may be found in ironstones and iron formations, but the carbonate layering has often been replaced by iron minerals.

The most interesting fact about iron formations and ironstones is that they have no modern counterpart. The only sedimentary iron ores forming today occur in lakes and bog areas and are nowhere near as large as the ancient iron-bearing deposits. This presents a puzzle that geologists have yet been unable to solve: Why are there no huge iron deposits forming today?

The iron in minerals occurs in two different forms: ferrous iron and ferric iron. The atoms of ferrous iron contain one more electron than the atoms of ferric iron. Ferrous iron is soluble in lakes, rivers, and seawater. Ferric iron is insoluble in surface waters. During present day weathering processes ferrous iron is generally converted to insoluble ferric iron through reactions with oxygen. Therefore, modern waters have a low concentration of iron in them. In order for the large iron deposits of the past to have formed, iron must have been mainly soluble ferrous iron. Many geologists believe this indicates that the earth's atmosphere more than about 570 million years ago contained much less oxygen than today. During weathering, little ferrous iron was converted to ferric iron. The absence of fossils in iron formations supports this conclusion. In fact, there are few fossils of oxygen-supported life in any rocks older than 570 million years.

Iron-bearing sedimentary rocks are a valuable resource. The iron used to manufacture steel comes from the ancient iron-rich sedimentary rocks. Without iron many of the items we take for granted would be unavailable. Buildings, appliances, plumbing—even pots and pans—would be vastly different.

Sedimentary rocks are very important economically. Their industrial uses in construction and as energy sources—coal and petroleum—vastly affect the world's cultures. The study of sedimentary rocks and processes changes almost every year as new laboratory techniques and equipment are invented. Continued study of sedimentary rocks and present day sedimentary processes will give us a more complete picture of these important rocks—what the earth was like long ago, and what it will be like in the future.

Glossary

algae—Plants without a true root, stem, or leaf that often grow in colonies in water.

aragonite—A calcium carbonate mineral that differs from calcite in crystal form, density, and the pattern in which it breaks.

arkose—A type of sandstone predominantly composed of feldspar and quartz.

breccia—A sedimentary rock composed of angular clastic pieces that have been cemented together.

calcite—A mineral composed of calcium carbonate found in limestone, chalk, and marble.

calcium carbonate—A chemical compound commonly found in calcite, aragonite, bones, shells, and some plant ashes.

carbonic acid—A weak acid.

chemical weathering—A weathering process that changes the chemical composition of a rock.

chert—A rock chiefly made of silica, that looks like flint.

clastics—Fragments of pre-existing rock.

coal—A combustible sedimentary rock formed from plant remains that have been partially decayed, buried, and changed as a result of the temperatures and pressure of burial.

coccoliths—The skeletal remains of a type of planktonic algae.

compaction—The pressing together of sediment by earth movements or overlying sediment layers.

conglomerate—Rounded pebbles and cobbles that have been cemented together.

cross-beds—Sediment layers deposited on an angle rather than horizontally.

erosion—The process of moving rock material to another area; water, wind, and ice often transport rock.

evaporite—Rock composed of minerals that precipitate as a solution evaporates.

feldspar—A rock forming mineral composed of silicon, oxygen, and aluminum.

formation—A sequence of rocks that from a unit that can be readily distinguished from rock above and below it.

gypsum—A soft evaporite mineral commonly found in sedimentary rocks.

igneous rock—Rock formed from cooled magma.

iron formation—Iron-bearing sedimentary rocks formed earlier than 570 million years ago.

ironstone—A sedimentary rock containing a high percentage of iron.

karst—An irregular limestone topography characterized by sinkholes and caverns.

lahar—A mudflow consisting of volcanic material, other rock and sediment, and water.

lamina—A very thick layer.

limestone—A sedimentary rock chiefly composed of the mineral calcite.

lithic sandstone—A type of sandstone characterized by angular pieces of quartz or feldspar that have been cemented together in a clay-size matrix.

lithification—The process by which loose rock material is solidified.

loess—Very small mineral fragments usually transported by wind.

metamorphic rock—A rock that has been altered from its original state by temperature and pressure.

micrite—Microscopic crystals of calcium carbonate.

mudrock—Sedimentary rock formed from silt-and clay-sized particles.

neomorphism—A change in the arrangement of atoms within sediment particles that occurs during lithification.

ooliths—Sand-size spheroidal grains, usually made of calcium carbonate.

ooze—A deposit of deep-sea mud.

parallel beds—Uniform sedimentary layering.

physical weathering—The breaking apart of existing rock without changing its chemical composition.

playa lake—A temporary lake formed in a desert drainage area.

precipitate—A substance that has separated and settled from a solution.

quartz—An important and very common rock-forming mineral composed of silicon and oxygen.

reducing environment—A stagnant, oxygen-free environment where little or no organic decay occurs.

reef—An underwater ridge of rock that extends to or near the water surface.

silica—A compound of silicon and oxygen commonly found in rocks as quartz.

sinkhole—A bowl-shaped depression found in regions underlain by limestone.

stromatolites—The laminated fossil remains of blue-green algae.

turbidity current—A current of cloudy or muddy water that moves because it is more dense than the surrounding water.

varve—A pair of thin sedimentary beds.

Further Reading

Barnes-Svarney, Patricia L. *Clocks in the Rocks: Learning About Earth's Past.* Hillside, N.J.: Enslow Publishers, Inc. 1990.

Bates, Robert L. *Stone, Clay, Glass: How Building Materials are Found and Used.* Hillside, N.J.: Enslow Publishers, Inc. 1986.

Bates, Robert L. *Industrial Minerals: How They Are Found and Used.* Hillside, N.J.: Enslow Publishers, Inc. 1988.

Deegan, Linda A. and Neill, Christopher. "New Life of the Mississippi." *Natural History*, June 1985, pp. 60-71.

Gould, Stephen Jay. "Treasures in a Taxonomic Wastebasket." *Natural History*, December 1985, pp. 22-33.

Pough, Frederick H. *A Field Guide to Rocks and Minerals,* 4th ed. (Peterson Field Guide Series.) Boston: Houghton Mifflin, 1976.

Rapp, George and Laura L. Erickson. *Earth's Chemical Clues: The Story of Geochemistry.* Hillside N.J.: Enslow Publishers, Inc. 1990.

Russell, George, reported by Bernard Diederich/Armero and Tom Quinn and Gavin Scott/Bogota. "Colombia's Mortal Agony." *Time*, November 25, 1985, p. 46.

Index